# Checkbook Math



REM 524

## A Teaching Resource From...

REMEDIA PUBLICATIONS

BLACKLINE MASTERS

To find Remedia products in a store near you, visit:
**http://www.rempub.com/stores**

**Remedia Publications, Inc.**
**15887 N. 76 Street • Suite 120 • Scottsdale, AZ 85260**

230

# INTRODUCTION

Of the many life skills students gradually learn, the handling of and accounting for money is an important survival objective. Early knowledge of the wisdom of saving and spending in a responsible way helps to establish good, life-long money-management habits.

*Checkbook Math* teaches the practical use of a checking account. It gives clear, simple instructions on how to write checks and includes lots of check-writing practice. Students also learn how to record checks and keep track of balances as well as how to add deposits to their accounts. Real-life situation word problems add reality and interest to this learning experience.

Reading: Grades 4-6 / Interest: Grades 6-12

# CONTENTS

# Parts of a Check

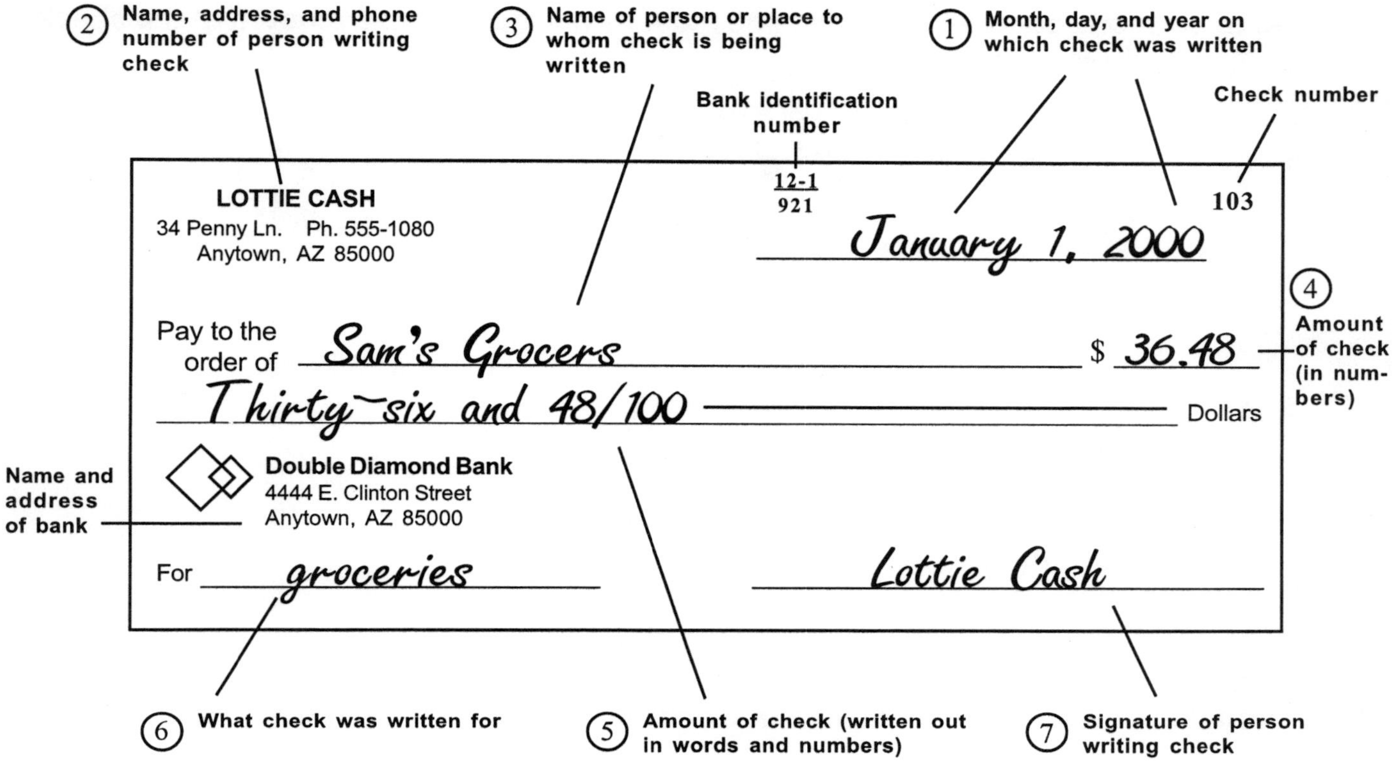


**Read the information given below. Then complete the blank check at the bottom of this page using this information.**

1. January 14, 2001
2. Print your name, address, and phone number here.
3. Toys 'R Fun
4. $26.75
5. Twenty-six and 75/100
6. birthday gift
7. Sign your name on this line.

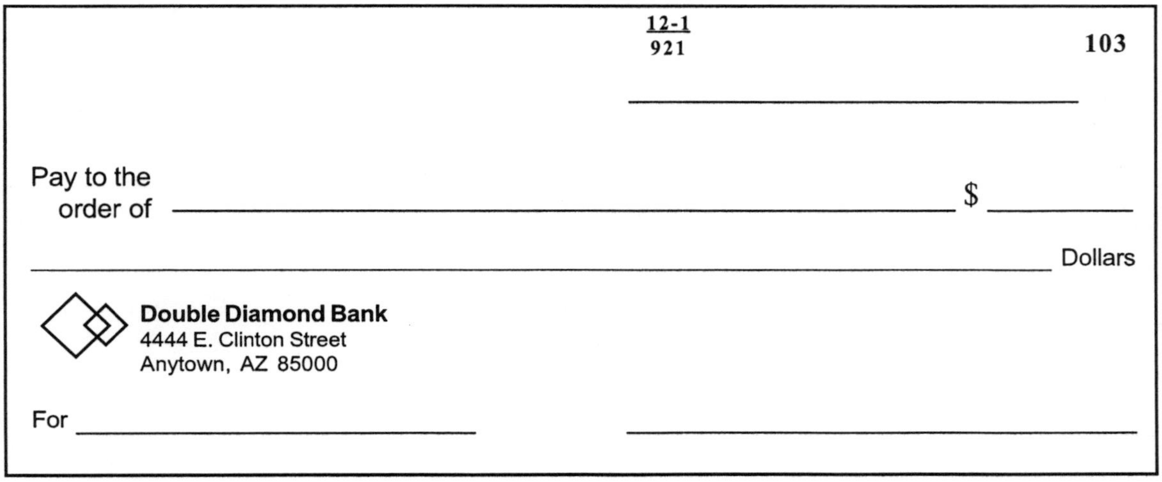

# Writing Dollars and Cents

**In order to correctly complete a check, you must be able to write the amount of the check in words. For example:**

Amount of check: **$125.36**

Dollar amount would be written as: ***One hundred twenty-five***

Cents would be written as: ***36/100***

**The amount of the check would then be written as follows:**

**One hundred twenty-five and 36/100** Dollars

**If there are no cents in the amount of the check, it is written as follows: 00/100 or no/100.**

**Read the different dollars and cents amounts below. On the line, write out these amounts as they should appear on the second line of a check. Use the glossary if you need help.**

1. $34.00 ______________________________
2. $96.00 ______________________________
3. $115.00

   ______________________________
4. $48.36 ______________________________
5. $27.18 ______________________________
6. $83.00 ______________________________
7. $12.50 ______________________________
8. $8.00 ______________________________
9. $126.54

   ______________________________
10. $210.00

    ______________________________

Name ______________________________



**Complete each check below with the information given. The name and address of the person writing the check has been printed on the check. Use this name to sign the check.**

| **Check No. 102:** | **Check No. 103:** | **Check No. 104:** |
|---|---|---|
| Date: July 16, 2001 | Date: February 13, 2001 | Date: April 4, 2001 |
| Paid to: Desert Flowers | Paid to: Sally's Bakery | Paid to: Jim's Nursery |
| Amount: $26.38 | Amount: $15.35 | Amount: $18.56 |
| For: Aunt Lillian | For: Valentine treats | For: plants, seeds |

**CARRIE SHORT**
4444 E. Tyler Ph. 555-1343
Phoenix, AZ 85000

12-1
921

102

______________

Pay to the order of ______________________ $ ________

______________________ Dollars

**Double Diamond Bank**
4444 E. Clinton Street
Anytown, AZ 85000

For ______________ ______________

**JAMES WOODS**
1212 N. Platt Ph. 555-1787
Anytown, AZ 85000

12-1
921

103

______________

Pay to the order of ______________________ $ ________

______________________ Dollars

**Double Diamond Bank**
4444 E. Clinton Street
Anytown, AZ 85000

For ______________ ______________

**BRENDA BLOOM**
3849 W. Rose Ph. 555-2666
Miami, FL 34000

12-1
921

104

______________

Pay to the order of ______________________ $ ________

______________________ Dollars

**Double Diamond Bank**
4444 E. Clinton Street
Anytown, AZ 85000

For ______________ ______________

 

**JEFF GOLDMAN**
1456 Silver St. Ph. 555-1189
Pittsburgh, PA 15000

12-1
921

102

______________________

Pay to the order of ______________________________ $ ________

______________________________________________ Dollars

**Double Diamond Bank**
4444 E. Clinton Street
Anytown, AZ 85000

For ____________________ ____________________

1. Jeff Goldman went to the Alpine Ski Shop to buy ski equipment. His total bill was $78.94. **Complete the check above to show Jeff's purchase. Use today's date.**

**SUSAN ALLEN**
352 Wood Ln. Ph. 555-3302
Little Rock, AR 51000

12-1
921

103

______________________

Pay to the order of ______________________________ $ ________

______________________________________________ Dollars

**Double Diamond Bank**
4444 E. Clinton Street
Anytown, AZ 85000

For ____________________ ____________________

2. Susan Allen bought a new outfit at Bloom's Department Store. She wrote a check for $43.52 on December 9, 2001, for her purchase. **Complete the check above to show this purchase.**

**ALLISON GREEN**
444 Lake Rd. Ph. 555-1818
Peoria, IL 60000

12-1
921

104

______________________

Pay to the order of ______________________________ $ ________

______________________________________________ Dollars

**Double Diamond Bank**
4444 E. Clinton Street
Anytown, AZ 85000

For ____________________ ____________________

3. Allison Green bought treats for her Girl Scout Troop. She made her purchases at Granny's Hometown Bakery. She wrote a check for $12.69 for the treats on February 8, 2001. **Complete the check.**

 

Name ______________________________



1. Sara Jane Benson is giving a birthday party for her friend, Kathy. At the Good Time Party Shop, she bought a tablecloth and napkins for $4.50, streamers and party hats for $2.85, and party favors for $12.00. Sara Jane wrote a check for the total amount. **Add Sara Jane's purchases together and complete the check below for the total amount. Use today's date.**

**SARA JANE BENSON**
1141 E. Thomas Ph. 555-3838
Phoenix, AZ 85000

12-1
921

102

______________________

Pay to the order of ______________________________ $ ________

______________________________________________ Dollars

**Double Diamond Bank**
4444 E. Clinton Street
Anytown, AZ 85000

For ____________________ ______________________

2. Sara Jane then went to Green's Corner Market. She purchased 3 cans of fruit punch for $3.29, two boxes of cake mix for $2.64, and two packages of frosting mix for $1.89. **Complete the check below to show the total amount of Sara Jane's purchases. Use today's date.**

**SARA JANE BENSON**
1141 E. Thomas Ph. 555-3838
Phoenix, AZ 85000

12-1
921

103

______________________

Pay to the order of ______________________________ $ ________

______________________________________________ Dollars

**Double Diamond Bank**
4444 E. Clinton Street
Anytown, AZ 85000

For ____________________ ______________________

**BONUS:** Sara Jane planned to spend $30.00 on the party. Did she stay within her budget?

__________ If so, how much does she have left? ______________________________

 

Kelly Jones and Jay French are having a Christmas party together.

1. Kelly went to ABC Hardware for lights and ornaments. She bought three packets of lights for $7.00, four boxes of balls for $12.36, and an extension cord for $2.29. **Complete the check below to show Kelly's total purchase. The date is December 10, 2001.**

**KELLY JONES**
490 E. Orange Ph. 555-4888
Tempe, AZ 85000

12-1
921

102

_______________

Pay to the order of ______________________________ $ ________

______________________________ Dollars

**Double Diamond Bank**
4444 E. Clinton Street
Anytown, AZ 85000

For ____________ ____________

2. On the same day, Jay bought a Christmas tree at Carl's Nursery for $21.50 and a tree stand for $6.85. He also purchased a can of spray snow for $2.35. **Complete the check below to show Jay's total purchase.**

**JAY FRENCH**
1180 N. Desert Ph. 555-4999
Tempe, AZ 85000

12-1
921

103

_______________

Pay to the order of ______________________________ $ ________

______________________________ Dollars

**Double Diamond Bank**
4444 E. Clinton Street
Anytown, AZ 85000

For ____________ ____________

**BONUS:** Kelly and Jay are sharing the cost of the party. Who has paid more so far?

______________________ How much more? ______________________

Name ______________________________________



1. Robert Gold is taking his family on a camping trip. At Four Peaks Sporting Goods, he bought a tent for $59.95, a camp stove for $24.95, and a large ice chest for $32.50. **Complete the check below for Mr. Gold's total purchase. Use today's date.**

**ROBERT & MARY GOLD**
8091 S. 10th St. Ph. 555-1004
Phoenix, AZ 85000

12-1
921

103

______________________

Pay to the order of ______________________________ $ ________

______________________________ Dollars

**Double Diamond Bank**
4444 E. Clinton Street
Anytown, AZ 85000

For ____________________ ______________________

2. Mrs. Gold went to Smitty's Grocery to buy some snacks for the trip. She bought two six packs of pop for $3.29, a bag of potato chips for $2.19, and a dozen donuts for $3.69. **Complete the check below that Mrs. Gold wrote for her purchases. Use today's date.**

**ROBERT & MARY GOLD**
8091 S. 10th St. Ph. 555-1004
Phoenix, AZ 85000

12-1
921

104

______________________

Pay to the order of ______________________________ $ ________

______________________________ Dollars

**Double Diamond Bank**
4444 E. Clinton Street
Anytown, AZ 85000

For ____________________ ______________________

**BONUS:** The Golds have budgeted $400 for their trip. How much money is left after their purchases?

________________ If they spend $50.00 on gas for the car, then how much will be left?

________________

Name ______________________________ 

1. Carl Lewis needed work done on his car. He went to Hal's Service Station. Carl had his engine tuned up for $29.95, bought a new battery for $39.95, and had the oil changed for $9.95. **Complete the check below to show Carl's total payment to the service station. Use today's date.**

CARL LEWIS
50 Glen Bay Dr. Ph. 555-1081
Mesa, AZ 85000

12-1
921

102

______________________

Pay to the order of ______________________________ $ __________

______________________________________________ Dollars

Double Diamond Bank
4444 E. Clinton Street
Anytown, AZ 85000

For ____________________ ______________________

2. Mary Alice Turner took her car to Swifty's Snappy Service. She had her engine tuned up for $64.00, bought a new battery for $49.95, and had the oil changed for $16.95. **Complete the check below to show Mary Alice's total purchase. Use today's date.**

MARY ALICE TURNER
1482 16th St. Ph. 555-4999
Phoenix, AZ 85000

12-1
921

103

______________________

Pay to the order of ______________________________ $ __________

______________________________________________ Dollars

Double Diamond Bank
4444 E. Clinton Street
Anytown, AZ 85000

For ____________________ ______________________

**BONUS:** If you needed work done on your car, would you take it to Hal or Swifty? Why?

______________________________________________________________

 

Name ________________________________



1. Michelle Winters went to the Shop-All store to stock up on some household items. She bought two packages of paper plates for $1.38, two bars of soap for 69 cents, two bottles of shampoo for $2.58, and two packages of trash bags for $5.00.

   She had discount coupons worth $1.05 that were deducted from her total. **Complete the check below to show Michelle's total purchase after her coupons were deducted. Use today's date.**

| | | |
|---|---|---|
| **MICHELLE WINTERS**<br>52 Sunny Lane Ph. 555-1404<br>Phoenix, AZ 85000 | 12-1/921<br>________________ | 103 |
| Pay to the order of ________________________ | | $ ________ |
| ________________________________________ | | Dollars |
| **Double Diamond Bank**<br>4444 E. Clinton Street<br>Anytown, AZ 85000 | | |
| For ________________ | ________________ | |

1. Michelle went to Larry's Drugs to make some additional purchases. She bought one bottle of aspirin for $3.49, a bottle of vitamins for $4.79, and a bottle of cough medicine for $2.49. **Complete the check below to show her total purchase. Use today's date.**

| | | |
|---|---|---|
| **MICHELLE WINTERS**<br>52 Sunny Lane Ph. 555-1404<br>Phoenix, AZ 85000 | 12-1/921<br>________________ | 104 |
| Pay to the order of ________________________ | | $ ________ |
| ________________________________________ | | Dollars |
| **Double Diamond Bank**<br>4444 E. Clinton Street<br>Anytown, AZ 85000 | | |
| For ________________ | ________________ | |

**BONUS:** The items Michelle purchased at Larry's Drugs were on sale. The regular price of the aspirin was $4.29; the vitamins, $5.35; and the cough medicine, $2.90. How much did she save by buying the items on sale?

________________

Name ______________________________



1. J.P. Nickel's is having their big summer sale. Billy Moore bought two shirts for $12.99 each and a new jacket for $39.00. **Complete the check below to show Billy's total purchase. Use today's date.**

| | | |
|---|---|---|
| **BILLY MOORE**<br>86 W. 59th Ave. Ph. 555-1208<br>Mesa, AZ 85000 | 12-1<br>921<br>______________ | 102 |
| Pay to the order of ______________________ | | $ ________ |
| ______________________________________ | | Dollars |
| **Double Diamond Bank**<br>4444 E. Clinton Street<br>Anytown, AZ 85000 | | |
| For ______________ | ______________ | |

2. Jerry Less went to J.P. Nickel's after the sale was over. He bought the same two shirts for $19.00 each and a new jacket for $52.00. **Complete the check below to show Jerry's total purchase. Use today's date.**

| | | |
|---|---|---|
| **JERRY LESS**<br>1003 Valley Dr. Ph. 555-1442<br>Phoenix, AZ 85000 | 12-1<br>921<br>______________ | 103 |
| Pay to the order of ______________________ | | $ ________ |
| ______________________________________ | | Dollars |
| **Double Diamond Bank**<br>4444 E. Clinton Street<br>Anytown, AZ 85000 | | |
| For ______________ | ______________ | |

**BONUS:** How much did Billy save on two shirts by buying them on sale? ______________

Name ______________________________



# Recording Your Checks

The amount of money in your checking account is called the **balance**. Each time you write a check, you must record it in your checkbook and **subtract** the amount of the check to get your **new balance**.

Mary Jones wrote the check below. Look at the check, and then look at how Mary recorded this check in her checkbook to get her new balance.

**Check:**

MARY JONES
306 S. Glover Ph. 545-2678
Phoenix, AZ 85000

12-1
921

104

July 7, 2001

Pay to the order of Lil's Beauty Shop $ 36.00

Thirty-six and 00/100 Dollars

Double Diamond Bank
4444 E. Clinton Street
Anytown, AZ 85000

For haircut

Mary Jones

**Checkbook:**

| Check Number | Date | Check Paid To | Check/Deposit Amount | Balance |
|---|---|---|---|---|
| | | | | 196.00 |
| | | | | −36.00 |
| 104 | 7/7/2001 | Lil's Beauty Salon | 36.00 | 160.00 |
| | | | | |
| | | | | |
| | | | | |
| | | | | |

1 Check number
2. Date check was written
3. To whom the check was written
4. Amount of the check
5. Balance before check was written
6. Amount of check written (subtract from balance)
7. New balance (amount left in checking account after check was written)

**Record these checks in the checkbook below and show the new balance.**

**Check No. 251:**
Date: 4/12/2001
Paid to: Larry's Dry Cleaners
Amount: $13.50

**Check No. 252:**
Date: 4/16/2001
Paid to: Millie's Fine Fabrics
Amount: $23.49

**Check No. 253:**
Date: 4/16/2001
Paid to: S & S Grocers
Amount: $47.56

| Check Number | Date | Check Paid To | Check/Deposit Amount | Balance 113.67 |
|---|---|---|---|---|
| | | | | |
| | | | | |
| | | | | |
| | | | | |
| | | | | |

**Record these checks in the checkbook below and show the new balance.**

**Check No. 185:**
Date: 7/9/2001
Paid to: Tom's Tires
Amount: $98.78

**Check No. 186:**
Date: 7/18/2001
Paid to: Flora's Fashions
Amount: $18.54

**Check No. 187:**
Date: 7/18/2001
Paid to: The Beautique
Amount: $36.00

| Check Number | Date | Check Paid To | Check/Deposit Amount | Balance 248.98 |
|---|---|---|---|---|
| | | | | |
| | | | | |
| | | | | |
| | | | | |
| | | | | |

 

Name ____________________



Philip Gray bought a stereo at Terry's Electronics for $149.95. Philip then went to Mel's Discount Music and purchased two new C.D.'s for $27.96.

**Complete both checks below to show Philip's purchases. Record the checks in his checkbook and show his new balance. Use today's date.**

**PHILIP GRAY**
15 Summit Pl. Ph. 555-4897
Tempe, AZ 85000

12-1
921

102

____________________

Pay to the order of ____________________ $ ________

____________________ Dollars

**Double Diamond Bank**
4444 E. Clinton Street
Anytown, AZ 85000

For ____________________ ____________________

**PHILIP GRAY**
15 Summit Pl. Ph. 555-4897
Tempe, AZ 85000

12-1
921

103

____________________

Pay to the order of ____________________ $ ________

____________________ Dollars

**Double Diamond Bank**
4444 E. Clinton Street
Anytown, AZ 85000

For ____________________ ____________________

| Check Number | Date | Check Paid To | Check/Deposit Amount | Balance |
|---|---|---|---|---|
| | | | | 225.00 |
| | | | | |
| | | | | |
| | | | | |
| | | | | |
| | | | | |

Name ______________________________



Terry Gordon was having some friends over to his house. He rented six movies at Vic's Video store for $12.39. Then he stopped at Gino's Pizzeria and bought three pizzas for $18.99.

**Complete both checks below and record them in his checkbook. Use today's date.**

**TERRY GORDON**
15 Sunset Dr. Ph. 555-8601
Scottsdale, AZ 85000

12-1
921

102

______________________

Pay to the order of ______________________________ $ ________

______________________________________ Dollars

**Double Diamond Bank**
4444 E. Clinton Street
Anytown, AZ 85000

For ____________________ ______________________

**TERRY GORDON**
15 Sunset Dr. Ph. 555-8601
Scottsdale, AZ 85000

12-1
921

103

______________________

Pay to the order of ______________________________ $ ________

______________________________________ Dollars

**Double Diamond Bank**
4444 E. Clinton Street
Anytown, AZ 85000

For ____________________ ______________________

| Check Number | Date | Check Paid To | Check/Deposit Amount | Balance 242.00 |
|---|---|---|---|---|
| | | | | |
| | | | | |
| | | | | |
| | | | | |
| | | | | |

**Bonus:** If Terry buys a case of pop for $12.50 and a large bag of popcorn for $2.69, how much will he have left in his checking account? ______________

 

Name ______________________________



At Arvin's Department Store, Cheryl Cross bought a bedspread for $69.95, two pillows for $15.00, and a set of sheets for $27.95. She then went to Jimbo's Paint & Tile and bought two gallons of paint for $25.98 and a roller and brush set for $9.95.

**Complete both checks to show Cheryl's purchases and record them in her checkbook. Use today's date.**

**CHERYL CROSS**
61 Green Rd. Ph. 555-0872
Phoenix, AZ 85000

12-1
921

102

_______________

Pay to the
order of ______________________________ $ ________

______________________________ Dollars

**Double Diamond Bank**
4444 E. Clinton Street
Anytown, AZ 85000

For ______________ ______________

**CHERYL CROSS**
61 Green Rd. Ph. 555-0872
Phoenix, AZ 85000

12-1
921

103

_______________

Pay to the
order of ______________________________ $ ________

______________________________ Dollars

**Double Diamond Bank**
4444 E. Clinton Street
Anytown, AZ 85000

For ______________ ______________

| Check Number | Date | Check Paid To | Check/Deposit Amount | Balance 213.47 |
|---|---|---|---|---|
| | | | | |
| | | | | |
| | | | | |
| | | | | |
| | | | | |

 

Name ____________________



Candy Crump started her Christmas shopping on November 23, 2001. At the Super Store, she bought a sweater for $28.00, a blouse for $13.99, two t-shirts for $12.00, and four pairs of socks for $10.00. At Terry's Toys, she bought a teddy bear for $22.95, a doll for $13.00, and a jigsaw puzzle for $4.99.

**Complete the checks and record them in her checkbook.**

CANDY CRUMP
15 South St. Ph. 555-8609
Mesa, AZ 85000

12-1
921

102

____________________

Pay to the order of ____________________ $ ________

____________________ Dollars

Double Diamond Bank
4444 E. Clinton Street
Anytown, AZ 85000

For ____________________ ____________________

CANDY CRUMP
15 South St. Ph. 555-8609
Mesa, AZ 85000

12-1
921

103

____________________

Pay to the order of ____________________ $ ________

____________________ Dollars

Double Diamond Bank
4444 E. Clinton Street
Anytown, AZ 85000

For ____________________ ____________________

| Check Number | Date | Check Paid To | Check/Deposit Amount | Balance |
|---|---|---|---|---|
| | | | | 136.72 |
| | | | | |
| | | | | |
| | | | | |
| | | | | |
| | | | | |

 

Name ______________________________



On August 22, 2001, Frankie Fishbomb went shopping at Bert's Books. He bought a dictionary for $12.95, a thesaurus for $6.95, and a book bag for $13.98. He then went to Slim's School Supplies. Here he bought one package of pencils for $1.19, and two spiral notebooks for $3.89.

**Complete Frankie's checks and record them in his checkbook.**

**FRANKIE FISHBOMB**
3 Trout Rd. Ph. 555-0018
Chandler, AZ 85000

12-1
921

**102**

______________________

Pay to the order of ______________________________ $ ________

______________________________ Dollars

**Double Diamond Bank**
4444 E. Clinton Street
Anytown, AZ 85000

For ____________________ ______________________

**FRANKIE FISHBOMB**
3 Trout Rd. Ph. 555-0018
Chandler, AZ 85000

12-1
921

**103**

______________________

Pay to the order of ______________________________ $ ________

______________________________ Dollars

**Double Diamond Bank**
4444 E. Clinton Street
Anytown, AZ 85000

For ____________________ ______________________

| Check Number | Date | Check Paid To | Check/Deposit Amount | Balance 64.37 |
|---|---|---|---|---|
| | | | | |
| | | | | |
| | | | | |
| | | | | |
| | | | | |

Name ____________________________________________



Billy Bucks has a lot of money. At Worldwide Travel, he bought an airline ticket to Hawaii for his friend, Darrell. The ticket cost $478.00. At Vic's Video, he bought a VCR for his pal, Larry. The VCR cost $229.36.

**Complete Billy's checks and record them in his checkbook. Use today's date.**

**BILLY BUCKS**
1 Golden Ln. Ph. 555-3333
Moneyville, AZ 85000

12-1
921

102

_______________

Pay to the order of ______________________ $ ________

______________________________ Dollars

**Double Diamond Bank**
4444 E. Clinton Street
Anytown, AZ 85000

For ______________ ______________

**BILLY BUCKS**
1 Golden Ln. Ph. 555-3333
Moneyville, AZ 85000

12-1
921

103

_______________

Pay to the order of ______________________ $ ________

______________________________ Dollars

**Double Diamond Bank**
4444 E. Clinton Street
Anytown, AZ 85000

For ______________ ______________

| Check Number | Date | Check Paid To | Check/Deposit Amount | Balance 2174.63 |
|---|---|---|---|---|
| | | | | |
| | | | | |
| | | | | |
| | | | | |
| | | | | |

 

Name ______________________________________________



Jodie Jensen went to the big sale at Teen Town. Here are the items she wanted to buy and the cost of each:

| | |
|---|---|
| 1 pair of jeans | $21.95 |
| 1 sweatshirt | $ 7.99 |
| 1 mini-skirt | $18.99 |
| 1 pair of boots | $18.00 |

Look at the balance in Jodie's checkbook. If she decides she *really* wants the jeans, what two items can she afford to buy with the jeans?

______________________________ ______________________________

**Complete Jodie's check for her purchase and record it in her checkbook. Use today's date.**

**JODIE JENSEN**
14 Curly Ln. Ph. 555-8882
Anytown, AZ 85000

12-1
921

104

______________________________

Pay to the order of ______________________________ $ __________

______________________________ Dollars

**Double Diamond Bank**
4444 E. Clinton Street
Anytown, AZ 85000

For ____________________ ____________________

| Check Number | Date | Check Paid To | Check/Deposit Amount | Balance |
|---|---|---|---|---|
| | | | | 48.00 |
| | | | | |
| | | | | |
| | | | | |
| | | | | |
| | | | | |

**Bonus:** How much more money does Jodie need to buy all four items? ______________

Name ____________________________________________



Phyllis and Frank Flake owe $67.86 to the Zapit Power Company, $34.83 to the Busy Bee Phone Company, and $44.56 to Sparkle Water.

They can only afford to pay two of these bills this week, but which two? **Help them out by completing the checks below and recording them in the Flakes' checkbook. Note: When two names are on a check, either person may sign the check.**

**PHYLLIS AND FRANK FLAKE**
36 Loopy Ln. Ph. 555-6867
Confusion, AZ 85000

12-1
921

**103**

______________________

Pay to the
order of ______________________________________ $ ________

__________________________________________________ Dollars

**Double Diamond Bank**
4444 E. Clinton Street
Anytown, AZ 85000

For ____________________ ______________________

**PHYLLIS AND FRANK FLAKE**
36 Loopy Ln. Ph. 555-6867
Confusion, AZ 85000

12-1
921

**104**

______________________

Pay to the
order of ______________________________________ $ ________

__________________________________________________ Dollars

**Double Diamond Bank**
4444 E. Clinton Street
Anytown, AZ 85000

For ____________________ ______________________

| Check Number | Date | Check Paid To | Check/Deposit Amount | Balance<br>*89.78* |
|---|---|---|---|---|
| | | | | |
| | | | | |
| | | | | |
| | | | | |
| | | | | |

 

Name ____________________



# Depositing Money in Your Checking Account

Before writing a check, you first must be sure you have money in your checking account.

You take money **out** of your account by **writing a check.**

You put money **into** your account by **making a deposit** at your bank.

Deposits need to be recorded in your checkbook, just like your checks. Instead of subtracting deposits from your balance, you **add** them to your balance.

The checkbook below shows the following checks and deposits. These are called ***transactions.***

1. Deposit made on 2/16/2001 for $74.00
2. Check written on 2/22/2001 for $34.00
3. Check written on 2/27/2001 for $6.84
4. Deposit made on 2/28/2001 for $55.00

| Check Number | Date | Check Paid To | Check/Deposit Amount | Balance |
|---|---|---|---|---|
| | | | | 124.69 |
| | 2/16/2001 | Deposit | 74.00 | |
| 115 | 2/22/2001 | Mac's Market | 34.00 | |
| 116 | 2/27/2001 | Daily News | 6.84 | |
| | 2/28/2001 | Deposit | 55.00 | |
| | | | | |

**Complete the checkbook to show the new balance after each transaction. Remember: SUBTRACT the checks from your balance; ADD the deposits to your balance.**

Gertie Gottaspend likes to shop. In one week, Gertie made all these transactions in her checkbook.

1. August 2: check #110 to Frilly Frocks for $34.56.
2. August 3: check #111 to Carla's Cookies for $12.50.
3. August 3: deposited $25.00 in her account.
4. August 4: received a check for $20.00 from Aunt Grace; deposited in bank.
5. August 5: check #112 to Top Ten Tunes for $14.79.
6. August 6: check #113 to Nick's Newsstand for $6.97.
7. August 6: check #114 to Doman's Department Store for $24.00.
8. August 7: Gertie received her paycheck; deposited $147.89 in her account.
9. August 8: check #115 to Fun Fashions for $67.00.

**Record all of Gertie's transactions in her checkbook. You will need to use two pages. Show Gertie's new balance after each transaction. Use this year.**

| Check Number | Date | Check Paid To | Check/Deposit Amount | Balance 168.50 |
|---|---|---|---|---|
| | | | | |
| | | | | |
| | | | | |
| | | | | |
| | | | | |

| Check Number | Date | Check Paid To | Check/Deposit Amount | Balance |
|---|---|---|---|---|
| | | | | |
| | | | | |
| | | | | |
| | | | | |
| | | | | |

 

Name ____________________



Susie Swatch received $75.00 for her birthday on March 20, 2001. She deposited this money in her checking account.

On March 22, 2001, Susie went shopping. She bought a new blouse at the Petite Palace for $14.99. She bought a new pair of shoes for $24.00 at the Boot Barn.

**Complete Susie's checks and record both her deposit and checks in her checkbook. Show her new balance after each transaction.**

**SUSIE SWATCH**
54 Cactus Ln. Ph. 555-3507
Globe, AZ 85000

12-1
921

**102**

____________________

Pay to the
order of ____________________ $ ________

____________________ Dollars

**Double Diamond Bank**
4444 E. Clinton Street
Anytown, AZ 85000

For ____________________ ____________________

**SUSIE SWATCH**
54 Cactus Ln. Ph. 555-3507
Globe, AZ 85000

12-1
921

**103**

____________________

Pay to the
order of ____________________ $ ________

____________________ Dollars

**Double Diamond Bank**
4444 E. Clinton Street
Anytown, AZ 85000

For ____________________ ____________________

| Check Number | Date | Check Paid To | Check/Deposit Amount | Balance 23.14 |
|---|---|---|---|---|
| | | | | |
| | | | | |
| | | | | |
| | | | | |
| | | | | |

Name ______________________________



Tubs Peterson has been working all summer doing errands for his neighbors. On his last job, the Wilsons paid him the following:

$10.00 for baby-sitting their dog
$ 5.00 for watering the lawn
$15.00 for mowing the grass
$ 4.00 for bringing in the newspapers
$12.00 for weeding the flower bed

Tubs put all but $10.00 of his earnings into his checking account.

Tubs now has enough money to buy a new bike at Willie's Wild Wheels. It costs $69.95. Tubs also wants to buy some reflectors for the bike. They cost $17.89.

**Record Tubs' deposit, made on August 10, 2001, in his checkbook, and complete the check he wrote, on August 11, 2001, to Willie's Wild Wheels.**

**TUBS PETERSON**
14 Elm St. Ph. 555-0697
Falling Oak, AZ 85000

12-1
921

**104**

______________________

Pay to the order of ______________________________ $ ________

______________________________________________ Dollars

**Double Diamond Bank**
4444 E. Clinton Street
Anytown, AZ 85000

For ____________________ ______________________

| Check Number | Date | Check Paid To | Check/Deposit Amount | Balance 60.00 |
|---|---|---|---|---|
| | | | | |
| | | | | |
| | | | | |
| | | | | |
| | | | | |

**Bonus:** Tubs receives $8 a week as allowance from his parents. How many weeks will he have to add his allowance to his new balance before he can afford to buy a skateboard for $49.95?

______________________________

 

Name ______________________________



Pretend the checkbook below belongs to you. Write one check to your neighborhood grocery store for $47.50. Write another check to your local department store for $67.80. Fill in the names of stores you have in your town. Use today's date.

Then make a deposit to your checking account in the amount of $136.50. **Complete the checks and the checkbook to show all these transactions. Be sure to include your name, address, phone number, and the dates you made the transactions. Show your new balance.**

12-1
921

102

______________________

Pay to the
order of ______________________________ $ ________

______________________________ Dollars

**Double Diamond Bank**
4444 E. Clinton Street
Anytown, AZ 85000

For ______________ ______________

12-1
921

103

______________________

Pay to the
order of ______________________________ $ ________

______________________________ Dollars

**Double Diamond Bank**
4444 E. Clinton Street
Anytown, AZ 85000

For ______________ ______________

| Check Number | Date | Check Paid To | Check/Deposit Amount | Balance 179.54 |
|---|---|---|---|---|
| | | | | |
| | | | | |
| | | | | |
| | | | | |
| | | | | |

## Writing Dollars and Cents on Checks

### — Glossary —

### Dollar Amounts

1: One
2: Two
3: Three
4: Four
5: Five
6: Six
7: Seven
8: Eight
9: Nine
10: Ten
11: Eleven
12: Twelve
13: Thirteen
14: Fourteen
15: Fifteen
16: Sixteen
17: Seventeen
18: Eighteen
19: Nineteen
20: Twenty

21: Twenty-one
22: Twenty-two
23: Twenty-three
24: Twenty-four
25: Twenty-five
26: Twenty-six
27: Twenty-seven
28: Twenty-eight
29: Twenty-nine
30: Thirty
40: Forty
50: Fifty
60: Sixty
70: Seventy
80: Eighty
90: Ninety
100: One hundred

101: One hundred one
102: One hundred two
103: One hundred three
104: One hundred four
105: One hundred five
106: One hundred six
107: One hundred seven
108: One hundred eight
109: One hundred nine
110: One hundred ten
111: One hundred eleven
112: One hundred twelve
113: One hundred thirteen
114: One hundred fourteen
115: One hundred fifteen
116: One hundred sixteen
117: One hundred seventeen
118: One hundred eighteen
119: One hundred nineteen
120: One hundred twenty

### Cents Amounts

0¢ 00/100 or no/100
1¢ 01/100
2¢ 02/100
3¢ 03/100
4¢ 04/100
5¢ 05/100

6¢ 06/100
7¢ 07/100
8¢ 08/100
9¢ 09/100
10¢ 10/100

15¢ 15/100
20¢ 20/100
25¢ 25/100
50¢ 50/100
99¢ 99/100

 

12-1
921

102

Pay to the order of ______________________ $ ________

______________________ Dollars

**Double Diamond Bank**
4444 E. Clinton Street
Anytown, AZ 85000

For ______________________ ______________________

12-1
921

103

Pay to the order of ______________________ $ ________

______________________ Dollars

**Double Diamond Bank**
4444 E. Clinton Street
Anytown, AZ 85000

For ______________________ ______________________

12-1
921

104

Pay to the order of ______________________ $ ________

______________________ Dollars

**Double Diamond Bank**
4444 E. Clinton Street
Anytown, AZ 85000

For ______________________ ______________________

| Check Number | Date | Check Paid To | Check/Deposit Amount | Balance |
|---|---|---|---|---|
| | | | | |
| | | | | |
| | | | | |
| | | | | |
| | | | | |

| Check Number | Date | Check Paid To | Check/Deposit Amount | Balance |
|---|---|---|---|---|
| | | | | |
| | | | | |
| | | | | |
| | | | | |
| | | | | |

| Check Number | Date | Check Paid To | Check/Deposit Amount | Balance |
|---|---|---|---|---|
| | | | | |
| | | | | |
| | | | | |
| | | | | |
| | | | | |

# Checkbook Math - Answer Key

**Page 2:** 1) Thirty-four 2) Ninety-six 3) One hundred fifteen 4) Forty-eight and 36/100 5) Twenty-seven and 18/100 6) Eighty-three 7) Twelve and 50/100 8) Eight 9) One hundred twenty-six and 54/100 10) Two hundred ten

| | | | |
|---|---|---|---|
| **Page 5:** | 1) Total purchase: $19.35 | 2) Total purchase: $7.82 | Bonus: Yes; $2.83 |
| **Page 6:** | 1) Total purchase: $ 21.65 | 2) Total purchase: $30.70 | Bonus: Jay; $9.05 |
| **Page 7:** | 1) Total purchase: $117.40 | 2) Total purchase: $9.17 | Bonus: $273.43; $223.43 |
| **Page 8:** | 1) Total purchase: $79.85 | 2) Total purchase: $130.90 | Bonus: Hal; He is cheaper. |
| **Page 9:** | 1) Total purchase: $8.60 | 2) Total purchase: $10.77 | Bonus: $1.77 |
| **Page 10:** | 1) Total purchase: $64.98 | 2) Total purchase: $90.00 | Bonus: $12.02 |

**Page 12:** First checkbook:

| | |
|---|---|
| Balance: | $113.67 |
| Check #251 | - 13.50 |
| Balance: | $100.17 |
| Check #252 | - 23.49 |
| Balance: | $ 76.68 |
| Check #253 | - 47.56 |
| Balance: | $ 29.12 |

Second Checkbook:

| | |
|---|---|
| Balance: | $248.98 |
| Check #185 | - 98.78 |
| Balance: | $150.20 |
| Check #186 | - 18.54 |
| Balance: | $131.66 |
| Check #187 | - 36.00 |
| Balance: | $ 95.66 |

**Page 13:**

| | |
|---|---|
| Balance: | $225.00 |
| Check #102 | - 149.95 |
| Balance: | $ 75.05 |
| Check #103 | - 27.96 |
| Balance: | $ 47.09 |

**Page 14:**

| | |
|---|---|
| Balance: | $242.00 |
| Check #103 | - 12.39 |
| Balance: | $229.61 |
| Check #104 | - 18.99 |
| Balance: | $210.62 |

Bonus: $195.43

**Page 15**

| | |
|---|---|
| Balance: | $213.47 |
| Check #102 | - 112.90 |
| Balance: | $100.57 |
| Check #103 | - 35.93 |
| Balance: | $ 64.64 |

**Page 16**

| | |
|---|---|
| Balance: | $136.72 |
| Check #102 | - 63.99 |
| Balance: | $ 72.73 |
| Check #103 | - 40.94 |
| Balance: | $ 31.79 |

**Page 17**

| | |
|---|---|
| Balance: | $ 64.37 |
| Check # 102 | - 33.88 |
| Balance | $ 30.49 |
| Check # 103 | - 5.08 |
| Balance: | $ 25.41 |

**Page 18:**

| | |
|---|---|
| Balance: | $2,174.63 |
| Check #102 | - 478.00 |
| Balance: | $1,696.63 |
| Check #103 | - 229.36 |
| Balance: | $1,467.27 |

**Page 19:** sweatshirt and boots

| | |
|---|---|
| Balance: | $ 48.00 |
| Check #104 | - 47.94 |
| Balance: | .06 |

Bonus: $18.93

**Page 20:**

| | |
|---|---|
| Balance: | $ 89.78 |
| Check #103 | - 34.83 |
| Balance: | $ 54.95 |
| Check #104 | - 44.56 |
| Balance: | $ 10.39 |

(phone & water)

**Page 21:**

| | |
|---|---|
| Balance: | $124.69 |
| Deposit | + 74.00 |
| Balance: | $198.69 |
| Check #115 | - 34.00 |
| Balance: | $164.69 |
| Check #116 | - 6.84 |
| Balance: | $157.85 |
| Deposit | + 55.00 |
| Balance: | $212.85 |

**Page 22:**

| | |
|---|---|
| Balance: | $168.50 |
| Check | - 34.56 |
| Balance: | $133.94 |
| Check | - 12.50 |
| Balance: | $121.44 |
| Deposit | + 25.00 |
| Balance: | $146.44 |
| Deposit | + 20.00 |
| Balance: | $166.44 |
| Check | - 14.79 |
| Balance: | $151.65 |
| Check | - 6.97 |
| Balance: | $144.68 |
| Check | - 24.00 |
| Balance: | $120.68 |
| Deposit | +147.89 |
| Balance: | $268.57 |
| Check | - 67.00 |
| Balance: | $201.57 |

**Page 23:**

| | |
|---|---|
| Balance: | $ 23.14 |
| Deposit | + 75.00 |
| Balance | $ 98.14 |
| Check #102 | - 14.99 |
| Balance: | $ 83.15 |
| Check #103 | - 24.00 |
| Balance: | $ 59.15 |

**Page 24:**

| | |
|---|---|
| Balance: | $ 60.00 |
| Deposit | + 36.00 |
| Balance: | $ 96.00 |
| Check #104 | - 87.84 |
| Balance: | $ 8.16 |

Bonus: Six